Gudgeon

T0079649

This book is a celebration of the l.
favourite 'tiddler'.

This much-loved little fish is long overdue a little book all of its own.

Scientist, author and broadcaster **Professor Mark Everard** tells tales about the biology of the gudgeon, gudgeon fishing and the diverse social quirks and values of this most popular of little fishes.

MARK EVERARD

Gudgeon
The Angler's Favourite Tiddler

CRC Press
Taylor & Francis Group
Boca Raton London

CRC Press is an imprint of the
Taylor & Francis Group, an **informa** business

Many thanks to David Miller for permission to use his wonderful gudgeon painting on the front cover
DAVID MILLER FISH & WILDLIFE ART
www.davidmillerart.co.uk

First edition published 2023
by CRC Press
6000 Broken Sound Parkway NW, Suite 300, Boca Raton, FL 33487–2742

and by CRC Press
4 Park Square, Milton Park, Abingdon, Oxon, OX14 4RN

CRC Press is an imprint of Taylor & Francis Group, LLC

Library of Congress Cataloging-in-Publication Data
Insert LoC Data here when available

ISBN: 978-1-032-31729-8 (hbk)
ISBN: 978-1-032-31727-4 (pbk)
ISBN: 978-1-003-31101-0 (ebk)

DOI: 10.1201/9781003311010

Typeset in Joanna MT
by Apex CoVantage, LLC

Contents

One

This book is about the gudgeon, a small fish familiar to most British anglers and other people interested in our fresh waters. Going by the Latin name Gobio gobio, the gudgeon is one of the smaller fishes commonly found in many British rivers and some still waters too.

Initially, perhaps appearing as a nondescript silvery fish, closer scrutiny reveals just what magnificent beasts gudgeon are. Closer examination of a gudgeon held up to the eye in good light is like looking into a prism, subtle violet shades suffusing a chain mail of conspicuous silvery scales. The flanks are adorned with brown and darker spots, iridescent

DOI: 10.1201/9781003311010-1

with multi-hued overtones, darker on the back and fading to creamy-white beneath. The body shape is gracefully streamlined. The rounded and disproportionately large head has a delicate underslung mouth fringed by a pair of short barbels. The eyes too are large. Subtle they may be, but gudgeon are ornate little mini-beasts!

Kenneth Mansfield wrote of the humble gudgeon in his delightful 1958 book *Small Fry and Bait Fish: How to Catch Them*:

> *Gudgeon are the most important fish mentioned in this book from the angler's point of view, for they provide sport, make very good live baits, serve the match fisherman well and are extremely good eating. Their nearest rival in these respects is the bleak.*

Gudgeon do not grow huge, nor are they rare. Despite that, they are almost universally liked, even loved, by anglers of all ages and those interested in the fishes of British fresh waters. Quite why they are so widely loved is a mystery . . . they just are!

This small book is a celebration of this most appealing of fishes and its many fascinating characteristics.

Two

Gudgeon, known by the Latin name *Gobio gobio* (Linnaeus, 1758), is a small fish widespread in the fresh waters of British, Europe and eastwards into Asia. Though small in stature, this is a fish that rouses disproportionately great interest and affection among anglers and naturalists.

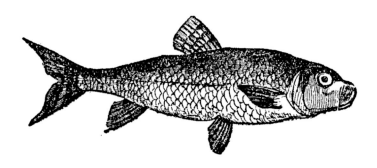

The gudgeon familiar to many in British waters are, as we will see in the following pages, far from the only fishes going by the name 'gudgeon' both in Europe and across the world.

GUDGEON TAXONOMY

Taxonomically, the gudgeon familiar to us in Britain are part of the class of ray-finned fishes (Actinopterygii), within which they are classified in the carp-like fishes (order Cypriniformes).

Within this broad order of carp-like fishes, the long-established family Cyprinidae (minnows or carps) covers some

DOI: 10.1201/9781003311010-2

3,160 species (in 376 genera) of varying forms occurring in fresh waters (only two species occur in fully marine waters) from North America, Africa and Eurasia. So diverse was this family of minnows and carps that it had been split into many quite distinctly different sub-families. One of these sub-families is (or was) the Gobioninae (the gudgeon).

However, over recent years, there has been growing consensus among scientists that this long-established family of fishes was too broad and that many of the sub-families it encompassed were, in reality, a grouping of quite distinct families. A 2018 reclassification of the carps and minnows, based on morphological and genetic differences, splits the large grouping into several distinct families. British examples of these newly redefined families include the revised Cyprinidae (true carps including common carp, barbel and crucian carp), Leuciscidae (minnows of Europe, Asia and North America including roach, common bream and Eurasian minnow), Tincidae (comprising the sole genus and species *Tinca tinca*, the tench) and Gobionidae (the gudgeon, including *G. gobio*).

The family Gobionidae, formerly the sub-family Gobioninae, now comprises 29 genera. Of the genus *Gobio*, only the familiar gudgeon *G. gobio* is native to and present in the British Isles.

KEY FEATURES OF THE GUDGEON

Gudgeon can grow up to 20 centimetres (8 inches) in body length, though they generally do not exceed 12–15 centimetres (5–6 inches). The maximum recorded weight of a gudgeon is 220 grams (slightly less than half-a-pound), though not from British waters. They are also recorded as living up to 8 years, though maximum ages of 6 or 7 are more common with most dying long before this. Females tend to be longer than males, which also tend to be less long-lived.

The gudgeon's body shape is elongated, tapering from a front half that is approximately round in cross section towards a laterally compressed tail. The head is heavy but short, and the eyes are conspicuous and large. The gudgeon's body is also slightly flattened beneath, betraying its bottom-dwelling habit.

The mouth of the gudgeon is 'inferior' (oriented downwards), adapted to its habit of feeding on the bed of the river or pool. The mouth has a single pair of barbels (short sensory whiskers), one at each corner. The number of barbels that this fish possesses is a useful diagnostic feature. It differentiates gudgeon from young barbel (*Barbus barbus*), which have two pairs of barbels surrounding their more robust underslung mouth as well as a smaller eye. Likewise, the number of barbels can distinguish small gudgeon from the two British loach species – the stone loach (*Barbatula barbatula*) and the spined loach (*Cobitis taenia*) – each of which possess three pairs of barbels. In common with other carp-like fishes, gudgeon have no teeth in the mouth, though teeth are present in the throat. These pharyngeal (or throat) teeth occur in two rows of 5 + 3 (or 2) on each side.

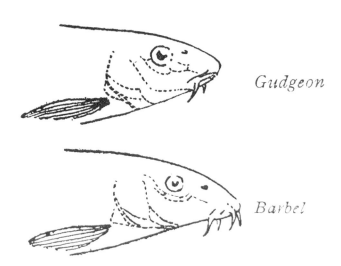

Gudgeon

Barbel

HEADS OF GUDGEON AND BARBEL

The body of the gudgeon is also covered evenly with large, conspicuous and pearlescent scales. The lateral line, a series of pits along the body containing organs sensitive to fluctuations in water pressure, is continuous along the flank spanning 38–44 of these scales.

Body colour varies enormously with habitat. Gudgeon from murky canals can often appear a mottled silvery with little more elaborate coloration. However, throughout most of their range, the back and sides are greeny-brown shading to silvery-yellow or creamy on the belly, and the flanks are adorned with large, indistinct dark mottling and a purplish, iridescent sheen. In fact, gudgeon can be veritable peacocks in terms of their subtle but dramatic body colour, fully matching the more overt splendour of fishes such as the grayling (*Thymallus thymallus*).

GUDGEON   BARBEL

The dorsal and tail (or caudal) fins are generally liberally spotted; the other fins are clear in appearance. The numbers of spines and soft, branched rays holding the fins erect are also used by fish biologists for accurate identification of species. The dorsal fin of the gudgeon is held erect by three spines (the first often very small) and five to seven soft branched rays (in scientific terms, this is described as III/5–7). The anal fin is III/6–7. Unlike barbels that have stout leading-edge spines to the dorsal and anal fins, those of the gudgeon are quite soft. Both the anal and dorsal fins of the gudgeon are short-based.

HABITAT PREFERENCES AND HABITS

Gudgeon occur in nearly all types of riverine and lacustrine habitats with a sandy or gravel bottom across their geographical range, though they may also proliferate in earthen-bedded waters too. They may be found in running waters, ranging from small mountain streams to larger lowland rivers. They also occur in large lakes and can proliferate also in canals. Though generally a freshwater species, gudgeon can also be found in some brackish habitats.

Gudgeon are benthopelagic, meaning that they inhabit the water just above the bottom. These fishes are generally active by day, though they can revert to nocturnal activity where predation pressures are intense.

In addition to their sense of sight, gudgeon also detect variations in water pressure through the lateral line system. They are also equipped with chemical sensors across the whole of the body surface. Gudgeon are also vocal, emitting squeaking vocalisations that are believed to be used to communicate with other gudgeon. Vocalisation varies with the degree of activity and temperature. As the vocalisations are independent of breeding season, they are thought not to be linked significantly to reproduction. Though possibly helping the fish shoal or alert each other to danger, the range of messages that gudgeon send to each other is not well understood.

Gudgeon prefer, and generally prosper best in, rapidly running water on sand and gravel (the 'grayling zone' of the river in some older river classifications). However, they are also found in slow-flowing lowland streams, and they often fare well in canals. They also occur in lakes with clear water and sand or stone bottoms, generally at shallow depth. During warmer months of the year, gudgeon tend to gather in small shoals in shallow water, whereas in winter, the fish retire to deeper waters.

Gudgeon also have a habit of periodically leaping clear of the water. While intuitively one might assume that this behaviour could be related to spawning or evading predators, gudgeon seem to do it outside of spawning periods as well as spontaneously when no predators are visible in the vicinity. Another theory is that fish of many species leap clear of the water to dislodge parasites, a more feasible explanation given that this behaviour seems widespread in most seasons and across many species. Also, the fact that gudgeon and other fishes tend to shimmy as they leave the water may support the theory of dislodging external parasites, though this may just be an artefact of them swimming rapidly upwards before breaking surface. Chris Yates recorded his fascinating observations about gudgeon in the famous Redmire Pool in Herefordshire, home of his and prior British Record common carp, in his classic 1986 book *Casting at the Sun*:

Another nice thing about the Redmire gudgeon was their tendency to buzz. On a still, summer night when you had been waiting by the rods for hours and nothing had stirred for all that time, it was amusing and somehow comforting to hear, close in, a 'gudgeon buzz'. The little fish, for a reason known only to itself, would leap right into the air and shake its body so violently that it vibrated like a bee's wing. But it was just a faint buzz, which was why it was audible only on still nights. For a time I presumed that carp-fry were responsible, but close observation in daylight revealed the true culprit. The small silver fish would pop up like a cork and then become just a blur before plopping back in the water. 'Bzzzzzzz!' it went.

GUDGEON DISTRIBUTION

Gudgeon are distributed widely across Europe. Though absent from Norway and the north of Sweden and Finland, they occur extensively southwards in catchments draining to the Atlantic Ocean, North and Baltic Sea basins, from the Loire eastwards, the Rhône and Volga drainages, upper Danube and middle and upper Dniestr and Dniepr drainages. Though naturally absent from Greece and the Iberian Peninsula, gudgeon now occur in both regions locally through introduction. Further to the east, gudgeon are found towards the former USSR and as far eastwards as Korea.

In the British Isles, gudgeon were naturally present only in the river catchments of Eastern England. This stems from a time up to the end of the last Ice Age (between 6500 and 6200 BC) when Britain was connected to continental Europe by a land mass known as 'Doggerland'. Doggerland today lies beneath the southern North Sea and is better known as Dogger Bank. When Britain was a peninsula of continental Europe, the Rivers Thames, Rhine and Scheldt formed the Channel River – today the English Channel – that carried their combined flow to the Atlantic. Gudgeon were present in the wider catchment of the Channel River, ranging from current British catchments from the Humber southwards to the Thames. They were restricted from spreading into river systems further to the west and north by impenetrable dry land. Towards the end of the Ice Age, a gigantic ice lake to the north broke to release a megatsunami that cut through and inundated the former Doggerland land connection, separating the British Isles from the continent.

Like many freshwater fish species, gudgeon have since been more widely spread by human activity from this former eastern British range. Gudgeon are now widely established across England apart from the far south west, though they do occur there locally. They now also occur locally in Scotland and Wales as a result of introductions, though they are increasingly scarce towards the north and west.

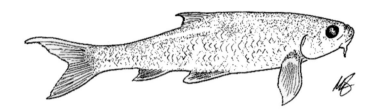

Gudgeon were also absent from Ireland. *Giraldus Cambrensis* ('Gerald of Wales', c.1146–c.1223), a medieval clergyman and chronicler of his times, published an account of *The History and Topography of Ireland*. In this book, he recorded that "*. . .pike, perch, roach, gardon, gudgeon, minnow, loach, bullheads and verones. . .*" were absent from Ireland. Giraldus observed that all the Irish species of freshwater fish known to him could live in salt water as migratory or brackish-tolerant species, listing brown trout, Atlantic salmon and arctic charr as well as pollan, three-spined sticklebacks, European eels, smelt, shad, three species of lamprey and the increasingly rare common sturgeon. These salt-tolerant fishes were all able to colonise Ireland's freshwater systems without man's interference, unlike many other fish species well suited to Ireland's diverse fresh waters. Common bream, roach, dace and tench have spread and thrived in Irish waters after introduction, and gudgeon too now occur in many parts of Ireland apart from the extreme west.

Gudgeon have also been introduced to eastern and northern Italy. However, their natural eastern and southern limits are unclear, as populations from the Iberian Peninsula and Adour basin in southern France as well as the Caspian Sea basin may be distinct species.

Other than by deliberate introductions for sport, food or ornamental purposes, some species are spread by accident. These accidental routes include as 'stowaways' with other stocked species, in bilge water or by invasion through human-made canals and water transfer schemes. The ramifications of some introductions are serious and addressed elsewhere.

The urban myth that fish eggs get carried to new waters attached to vegetation adhering to the feet of ducks is totally lacking in supporting scientific evidence. However, a study published in 2020 showed that a proportion of live embryos could be retrieved, and subsequently hatched, from the faeces of captive mallard ducks fed with feed that included the developing eggs of two species of carp (common carp and Prussian carp). Quite how this newly proven phenomenon known as 'endozoochory' relates to the gudgeon as a means of distribution is uncertain. However, as the eggs of gudgeon are generally thought to be released over sandy beds, to which they then fall and stick, they are far less likely to be ingested by ducks grazing on submerged vegetation.

THE GUDGEON DIET

Gudgeon are a bottom-dwelling and bottom-feeding species. Their diet is truly catholic. They feed opportunistically and

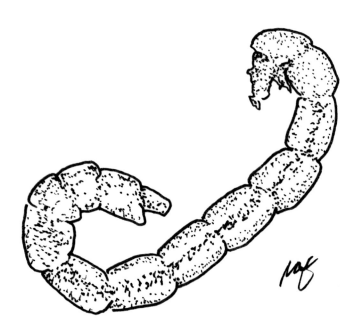

omnivorously on the bed of the waterbody, seeking out the larvae of insects, including bloodworms (the bright red larvae of chironomid midges that can form dense beds in muddy bottoms), worms, crustaceans, molluscs and, occasionally, fish eggs and even fish fry as well as small pieces of vegetation and other organic matter.

These small food items are sucked in through the protrusible mouth. Gudgeon feed all year, in winter and summer alike.

REPRODUCTION AND LIFE CYCLE

Gudgeon spawn communally in shallow water over stones, sand or plant material between mid-April and sometimes even as late as early July. This generally occurs when the water reaches 15°C. At this time, male gudgeon develop a dense covering of spawning tubercles (wart-like projections), particularly on the head and the front of the body that are thought to stimulate egg release by female fishes.

Gudgeon are commonly described as psamophilic spawners, meaning that they prefer to spawn over sand. Frequency of spawning varies with the richness of the water. In low-productivity streams, spawning may occur only once per year, whereas multiple spawnings may occur in environments that are richer in nutrients and so higher in productivity.

Females shed between 1,000 and 3,000 eggs at intervals in batches of 50 or 100 over several days. The eggs are fertilised by milt from male fish, released as female gudgeon drop their eggs. Gudgeon eggs measure 1.5–2 millimetres in diameter and are translucent with a mauve or bluish tinge. The eggs are sticky, released above the substrate to drift with the current before sinking to the bottom and adhering to the bed.

A. Lawrence Wells describes the fate after release and the appearance of gudgeon eggs in his 1941 book *The Observer's Book of Freshwater Fishes of the British Isles*:

> *Sinking to the bottom and there adhering to the stones, so that the flow of water will not wash them away, they may be seen as the sun glints across them like tiny soap bubbles.*

However, my friend and colleague Dr Adrian Pinder, the author of the definitive 2001 publication *Keys to Larval and Juvenile Stages of Coarse Fishes from Fresh Waters in the British Isles*, has some interesting observations about these alleged psamophilic spawning habits. In conducting research underpinning his keys, Adrian collected fish eggs from a range of habitats to grow on in captivity, drawing and photographing life stages and determining what species they were as they grew.

Adrian says that he never found gudgeon eggs on sand, though he acknowledges that such scattered and translucent eggs would be hard to spot there. However, he did harvest eggs from *Fontinalis* moss growing in very atypical gudgeon-like shallow fast water on spill weirs on the Great Ouse. Initially unidentified, these eggs were hatched in the laboratory, and they grew on to become gudgeon. Perhaps understandably, little resource is given to studying our smaller fish species. Consequently, many assumptions are made and repeated often enough to become received truths, yet these may perhaps be only part of the full reality.

Gudgeon exhibit no parental protection of the eggs, which are consequently heavily predated by a wide range of fish and invertebrates. Surviving fertilised eggs hatch after 10–30 days, depending on the temperature of the water. When first hatched, the larval fish are incompletely developed and remain immobile while consuming their attached yolk sac. After the yolk is consumed, the young become free-living and shoal at the bottom near the spawning location. Larvae and juveniles occur on the bottom, preferring detritus-rich sandy habitats and low current speeds.

Subsequent growth rate varies widely from water to water, food availability and temperature exerting significant influences. Initial growth in the first two years is rapid, with juvenile gudgeon often growing by as much as 5–6 centimetres (2 inches) per year. Growth slows beyond this initial surge.

Both male and female gudgeon generally mature in their third year at an average length of 10 centimetres (4 inches). However, some may mature in their second year. Gudgeon are not long-lived, generally reaching a maximum life span of 6 or 7 years.

PREDATORS OF GUDGEON

In common with many small fishes of British freshwaters, gudgeon are predated on by many other animals. These predators include various fish species such as perch, pike, zander, chub and eels. As Mr and Mrs S.C. Hall relate the perch and the gudgeon in their 1859 book *The Book of the Thames From Its Rise to Its Fall*, the gudgeon

> . . . is commonly found where the perch luxuriates; although associates, however, they are by no means friends – on the contrary, the one is the prey of the other.

Many other omnivorous fishes will not turn down a small gudgeon that swims their way.

A wide variety of piscivorous birds – herons, egrets, cormorants and kingfishers, among them – also readily take gudgeon. Mammals too can take gudgeon, including otters whose foraging tactics on the beds of rivers make bottom-dwelling species such as bullheads, eels, loach and gudgeon ideal prey. However, otters have a stone-turning feeding habit, and these are the habitats of bullheads, loach and eels. By contrast, gudgeon are fishes of open river beds, so they may not be as readily caught.

Another amazing and surprising predatory event was narrated by the Reverend Victor James in his essay *Fishing Shrews* (compiled into BB's *The Fisherman's Bedside Book*). The Reverend James makes a curious observation about the mysterious disappearance of gudgeon and other small fishes that he had stocked into a stream-fed water butt and a deepened stream channel in his garden:

> *The solution to the problem offered itself most unexpectedly in the following manner: One afternoon I was looking into the water-butt mentioned before, where I had stocked a reserve of roach and gudgeon, when I saw a quick movement opposite. A shrewmouse appeared from inside the inlet pipe, poised itself of the edge, sniffing the water delicately, then dived, using its large forefeet with extraordinary rapidity to assist its downward progress. For a matter of seconds it swam at the bottom, then rose, and, evidently alarmed by some slight movement of surprise on my part, it disappeared into the pipe whence it had previously emerged.*
>
> *I had a long net near at hand and used it at once. After one or two dips the net brought up a gudgeon newly severed in two!*

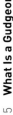

Following such a revolutionary discovery we kept close watch on the small stream and were rewarded by seeing a shrew find an easy prey in the shallower water. It seized a gudgeon and severed it in half as before.

Subsequently we found the remains of minnows, roach, and even perch that had been similarly treated. Needless to say, we removed the rest of our defenceless friends from danger.

OTHER 'GUDGEON'

Other species of fish beyond this range go by the name 'gudgeon'. For example, some 40–50 species of freshwater fishes distributed across Australia are known as 'gudgeon'. These include some fish up to 3 kilograms, though most are far smaller, including various attractive aquarium species. However, these fish are in the goby family (*Gobiidae*), quite unrelated to the European gudgeon.

Another 'gudgeon' that is not a true gudgeon is the topmouth gudgeon (*Pseudorasbora parva*), a member of the minnow and carp family also known in the aquatic trade as 'stone moroko' or 'clicker barb'. Topmouth gudgeon are small fish weighing up to 16 grams (half an ounce), native to flowing and still fresh waters from Japan to the Amur basin. However, they now are found in some British waters as a problematic alien invasive species, probably spread from the aquarist trade. Though small, topmouth gudgeon have a rapid growth rate, the capacity to breed after only their first year, and can spawn multiple times each year on leaves or stones throughout the spring and summer. Male topmouth gudgeon also exhibit a degree of parental protection of these eggs. Feeding on invertebrates, fish fry and fish eggs, topmouth gudgeon can rapidly colonise and proliferate in new waters and, due to their tendency to eat the spawn and fry of other fishes, both outcompete and prevent the reproduction of native species. Active eradication programmes are in place to eliminate topmouth gudgeon from British waters.

Topmouth gudgeon are far from the only problematic 'alien invasive species' found in British waters. Species of plants and animals introduced beyond the natural range within which they co-evolved are a cause of major concern globally.

MORE INFORMATION ABOUT GUDGEON

If you want to know more details about the biology of gudgeon and other British freshwater fishes, I can point you towards three of my other books:

- *The Complex Lives of British Freshwater Fishes.* (2020). CRC/ Taylor and Francis, Boca Raton and London.
- Everard, M. (2013). *Britain's Freshwater Fishes*. Princeton University Press/WildGUIDES, Woodstock, Oxfordshire.
- Everard, M. (2008). *The Little Book of Little Fishes*. Medlar Press, Ellesmere.

Three

Arthur P. Bell, in his 1926 book *Fresh-Water Fishing for the Beginner*, described the gudgeon affectionately as 'A SPORTING little river fish'. Many writers underline the fact that there are no special tactics for gudgeon fishing, unlike advice given for pursuing many larger fish species. However, there are some general principles and a few tricks that can increase your success when you go out fishing for gudgeon.

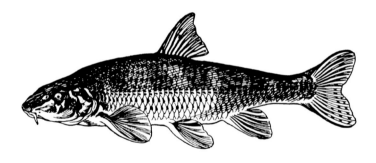

Like many anglers growing up in the 1960s, the advice of the great pioneering angler Richard is deeply ingrained in my pursuit of fish of all species.

Stuart Walker is deeply ingrained in my pursuit of fish of all species. Until the 1950s and 1960s, anglers generally assumed that catching a big fish was a matter of pure chance, with the landing of a specimen a rare and unpredictable occurrence. Walker scotched that myth through his innovative, scientific approach, astounding tally of record and specimen coarse and game fishes, and his clear and prolific writings. At the heart of

DOI: 10.1201/9781003311010-3

Walker's approach was what I consider a 'holy triad' of steps: location, bait and presentation.

Location is obviously vital, as you simply can't catch fish that are not there, or at least that are not feeding. Then, you have to match what they are feeding on, or else induce them into taking what you intend to offer as bait. And, of course, presentation is a matter of choosing the right tactics to offer the bait in the right place and at the right time without arousing suspicion from the fish, and then to be able to detect bites and, hopefully, subsequently land the fish.

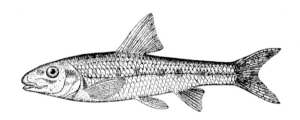

GUDGEON LOCATION

Gudgeon are opportunistic fishes. Although often described as a river fish, they are nonetheless adapted to a wider variety of freshwater and also some brackish habitats. Gudgeon commonly prosper in canals, as well as in some lakes and ponds, and can be prolific in food-rich waters such as commercial still water fisheries.

In rivers, food-rich shallow areas with a flow of water are favoured. For example, Mr Lane, in his 1843 book *The Diary of A. J. Lane: With a Description of Those Fishes to be Found in British Fresh Waters*, advised the gudgeon angler:

> *Go to shallows and gravelly places free from weeds or where cattle have gone in to drink.*

In faster rivers, slower margins, backwaters and back eddies off the main flow can be productive areas for finding gudgeon.

The advice that gudgeon are attracted to a disturbed bed also holds true.

A technique touched upon by many older angling writers is that of raking the river bed to attract gudgeon and stimulate them into feeding. This is not such an odd concept. After all, we know that robins learn to follow truffling wild boars or gardeners turning the soil, rushing in to snaffle worms and other small soil creatures that their actions expose. Also, that gulls learn to follow a plough for the same reason. When feeding in earnest on the bed of a river, gudgeon may sometimes be seen 'flashing', as they turn on their sides to disturb the river bed in pursuit of small food items. It is consequently unsurprising that they are attracted to bed disturbance.

Advice on raking the river bed to attract gudgeon dates back centuries. For example, it is written about by John Dennys, believed to be a fishing companion of William Shakespeare, in his 1613 book *The Secrets of Angling*.

In *The Diary of A. J. Lane: With a Description of those Fishes to be found in British Fresh Waters*, Mr Lane writes:

In the Thames they always use a raik [sic] (and by raiking the ground and discoloring the water occasionally) draw them together. These lively little fish afford much sport and amusement. 25 or 30 dozen may be taken with one rod in a day and I have known of 400 being taken at Henly [sic]. I never myself caught more than 27 dozen in a day.

Other angling writers offer similar advice. For example, in his 1943 book *Coarse Fish*, Eric Marshall-Hardy, one-time editor of *Angling* magazine, wrote:

If the angler will take the trouble to rake the bottom regularly while fishing they may be caught by the hundred and constitute very delicate and appetizing food, not to speak of their usefulness as live baits for Pike, Perch and Chub (the smallest for the latter).

Also, in *The Observer's Book of Freshwater Fishes of the British Isles*, published in 1941, A. Lawrence Wells recorded:

The boat-keepers and fishermen of the Thames use a large iron rake which they throw overboard and so rake the gravel; 'scratching their backs,' as they say. The stirred up gravel will soon attract the fish, but, unless the bait is at just the right distance from the bottom you may not catch a fish even though hundreds of them are playing round it.

In his 1958 book *Small Fry and Bait Fish: How to Catch Them*, Kenneth Mansfield wrote:

Once a shoal of gudgeon has been located, by observation or by trial-and-error angling, rake the bottom.

Arthur P. Bell added in his 1926 book *Fresh-Water Fishing for the Beginner*:

If possible rake the bottom of the river if it is at all gravelly, and the sport will be fast and furious until Master Perch looks in, and then there is generally a smash.

Kenneth Mansfield also noted that raking of the river bed was, as it still is, banned in most matches.

Many a pleasure, match and younger anglers will also have learned that a little disturbance of the bed of the river or pool can aid their pursuit of gudgeon and other small fishes. Whether by gently trampling with waders, wellington boots or sandals in shallow water, or with the tip of a rod, pole or landing net handle in deeper margins, a little disturbance of the bed to kick up sediment to cloud the water and expose small creatures can bring in the desired small fishes. If gudgeon are present, they may home in on this disturbance and can be fished for almost immediately. An added bonus is that other fish may also be attracted to the puff of sediment and small creatures released not only by the stirring but also by the activity of gudgeon foraging for loose morsels. This, in microcosm, is the modern and less invasive version of raking in action, as raking itself is inconsistent with some match-fishing rules and there is also a more general need to avoid excessive damage to the fishery.

Another effective and more flexible approach not so much to locating gudgeon as to localising them where they are present is liberal groundbaiting. A base mix of groundbait will add some cloud and flavour, and ideally will have mixed in with it some of the bait (such as maggots or worm fragments) that the angler intends to use to catch gudgeon. An important consideration here is that the groundbait has to reach the bed of the river or pool and not be swept away by currents. This either entails a heavy mix that sinks quickly or a targeted delivery method such as a bait dropper or a swimfeeder.

Gudgeon seem comfortable in all depths of water, from the margins of lakes and rivers down to substantial depths, but the key consideration is to keep the bait on or near the bed. This may be problematic in large still waters in the summer, as lower layers may become stratified and low in oxygen, in which case the gudgeon will move onto shallower margins and gravel bars.

Gudgeon are also reliable feeders in most weather, even in cold winter conditions. This makes them a dependable source of sport, a target for match anglers in adverse conditions and a resource as a bait for predator fishing. However, distribution in any waterbody tends to vary between summer and winter. Gudgeon tend towards streamier runs of running water or the margins of lakes and canals in warmer conditions. However, they head to deeper still and flowing waters in winter. As Izaak Walton wrote in *The Compleat Angler*:

> They be usually scattered up and down every river in the shallows, in the heat of summer: but in autumn, when the weeds begin to grow sour and rot, and the weather colder, then they gather together, and get into the deeper parts of the water . . .

TRENDS IN GUDGEON DISTRIBUTION

I was asked several times when writing this book why gudgeon seemed to be scarce where once they were prolific and, conversely, why these little fishes were proliferating in other places where they had not before. In a less than scientific survey, I asked a range of fishery managers and angling friends from across the country, backed up by queries on social media. What then did I learn?

On some rivers formerly holding many gudgeon, including big gudgeon – most notably the River Nadder where the current British Record was caught as well as the Dorset Frome – these fishes seemed, at the time of writing, to have become almost absent. Similar tales are told of other once-prolific venues ranging from the Rivers Trent, Wensum and Lea, and canals including the Trent and Mersey, Basingstoke, Shropshire, Erewash and Grand Union.

By contrast, I was told that that gudgeon were to be found 'one a chuck' in various other canals. My own experience endorses that of others concerning the current abundance of gudgeon in the Kennet and Avon Canal. I was also told that a large number of still waters across the country have thriving gudgeon populations, including some quite big specimens occurring in commercial fisheries. Some rivers also seem to be seeing good stocks of gudgeon, including the Loddon, the Wandle and other tributaries of the wider Thames catchment as well as the River Severn and some of its tributaries such as the Stour. The River Erewash is also said to hold large gudgeon (the Erewash is a tributary of the Trent noted elsewhere for its decreasing gudgeon population). In the east of England, I am told that the Bure holds a strong gudgeon population.

This is a just a snapshot of a moment in time, based on wide-spread but unscientific sampling. Other than the increasing number of commercial fisheries that did not previously exist hosting prolific stocks of gudgeon, there is no clearly evident pattern. As a generality, places with 'perfect' gudgeon habitat – gravel and sandy bottoms with good water quality – seem often to be suffering. Conversely, waters with ostensibly less ideal conditions, such as muddy pits and canals, appear to be doing rather well! But there are enough waters currently bucking those apparent trends, such as the reportedly good stock and large sizes of gudgeon on the River Severn, to discount these general observations!

This, of course, leads us to ask why these trends are happening. Undoubtedly, agricultural intensification is resulting in increasing nutrient-rich and silted rivers, and this could have a negative effect. So too does increased predation, for example through the spread of zander and cormorants, though many reported declines precede their current prevalence. The spread of common carp and American signal crayfish, both opportunistic feeders including on fish eggs and small fish and both with additional wider and often significant impacts on habitat, could also be explanatory. Yet none of these factors alone seems satisfactorily to explain the upward and downward trends we have seen, and doubtless will continue to see, in gudgeon populations.

A further interesting reflection is that angling methods have also changed over the years, with a general trend towards larger baits, hooks and stronger lines targeting bigger species of fish. There has also been a substantial decline in club match angling across the country with a corresponding reduced focus on catching gudgeon, match angling lurching towards commercial pits generally targeting carp and other larger species. Changing angling habits then also compound possible explanations of why fewer gudgeon are caught where they were once known to be present.

Where to go then to locate the biggest specimen gudgeon? My money would still be on a river with good water quality as well as a clean sand and gravel bed, such as the Hampshire Avon, Severn or Test, even if the head of gudgeon there is slighter than once it was. But, if I was intent on a record – I am not! – I would follow advice I have given before regarding other species. That is to target commercial fisheries that see a huge amount of bait, upon which neglected opportunistic species such as roach and gudgeon can grow fat and unnoticed!

GUDGEON BAITS

As we learned when considering the catholic tastes of this bottom-dwelling and bottom-feeding species, gudgeon feed opportunistically on a range of small items including both animal and other organic matter on or near the bed.

The larvae of insects are important in the gudgeon's diet. Maggots of all types – pinkies, squats, maggots dyed in various colours, gozzers and even casters – make fine gudgeon baits in all waters. Some match anglers favour dead maggots over live ones.

Bloodworms are also accepted by gudgeon, though procuring and presenting them are a little more specialised and potentially expensive than may be necessary for this less than fussy fish. Worms too are a fine gudgeon bait, particularly smaller worm species such as dendrobaena and brandlings, or fragments of them. Where you can easily procure freshwater shrimps, they will be taken, though there are easier baits to find and use including, for example, fragments of prawns that may be bought from chiller cabinets in most supermarkets. But other animal-based parts of the gudgeon's diet — molluscs and occasionally fish eggs and even fish fry — demonstrate what opportunists these little fishes are, and therefore that any suitable small fragments of bait may be found and consumed if presented correctly.

The 1496 *Treatyse of Fysshynge with an Angle*, credited to the English nun and writer Dame Juliana Berners, offers the observation:

> The Gogen is a good fysshe of the mochenes [for its size]; and he biteth wel at the grounde. And his baytes for all the yere ben thyse: ye red worme: codworme [caddis]: and maggdes [maggots].

Many other angling writers describe their favourite baits for gudgeon fishing. One such is Mr Lane, who, in his 1843 book *The Diary of A. J. Lane: With a Description of Those Fishes to be Found in British Fresh Waters*, commends:

> Bait red worms or gentles. [As many older angers will remember, 'gentles' was a common name for maggots at least through to the 1960s.]

Henry Coxon, in his 1896 book *A Modern Treatise on Practical Coarse Fish Angling*, comments:

> They are fond of gravelly bottoms and are easily caught, either with a bit of a worm, or a gentle.

Gudgeon appear to have a distinct preference for meaty baits in warmer weather. Finely diced luncheon meat cubes can be extremely effective in warm seasons, not to mention convenient. Apparently, pieces of minced steak can also attract hungry gudgeon. Also, in the early 1960s, I caught them on wasp grubs, though these grubs are no longer a commonly available bait today.

Gudgeon also feed on vegetable and other amorphous organic matter such as detritus, a nutritious food owing to the proliferation of microbes within it. I catch many of them while roach fishing using breadflake, punched bread or bread paste. Other types of pastes or fragments of boilies are also keenly consumed. In fact, few baits of a suitable size will be refused by gudgeon.

Groundbait to attract gudgeon need not be complex or expensive. However, some anglers note that gudgeon have a taste for sweet and spicy additives, including molasses, turmeric or cinnamon. Experimentation may reap rewards, though I have not found that anything more complicated than liquidised or mashed bread, laced if necessary with maggots or worm fragments if I am using them as bait, can yield success in terms of creating a mist of fine particles and offerings of small food items to whet their appetites.

PRESENTATION WHEN GUDGEON FISHING

As we observed when reviewing their ecology, gudgeon are both bottom-dwelling and bottom-feeding. This is an important consideration, as a bait, any significant distance above the bed of the waterbody may not be intercepted no matter how many gudgeon are present.

Beyond this constraint, tackle choice need not be specialised. As Kenneth Mansfield put it in his 1958 book *Small Fry and Bait Fish: How to Catch Them*:

> It is unnecessary to go into long details of tackle and methods.

On this point, I have to agree, as gudgeon can be taken on many methods provided, they present the suitable bait on the bed of the river, canal or pool. Mansfield expresses the opinion:

> Float fishing is the only method worthy of consideration.

Though a fan of the float, I am only 50% in agreement as leger tactics – considerably refined since the 1950s – catch many a gudgeon too!

The key consideration is to present a suitable bait on or near the bed of the river, canal or lake. As Izaak Walton put it in *The Compleat Angler* regarding gudgeon fishing:

> He and the Barbel both feed so: and do not hunt for flies at any time, as most other fishes do. He is an excellent fish to enter a young angler, being easy to be taken with a small red worm, on or very near to the ground.

Walton also adds that gudgeon

are to be fished for there, with your hook always touching the ground, if you fish for him with a float or with a cork.

This brings us on to the topic of float fishing. Stick float fishing is generally the best bet in running waters for gudgeon. Stick float size and weight need to be matched to flow conditions, generally with shot strung evenly up the line 'shirt button style'. However, in heavier flows, a larger Avon or Loafer style of float may be necessary. When using a larger float of this type in running water, a bulk weight between float and hook link will take the bait down more quickly in the water column to the feeding depth. For sensitive presentation, a 'dropper shot' is placed between the bulk weight and the hook length to bring the bait closer to the river bed in a subtle manner, such that the hook and chosen bait can trip along gently where the fish are believed to be feeding.

In still or slow-moving water, waggler float tactics may work if the water is shallow. The key point is that the bait needs to trip the bottom where the gudgeon can be expected to be feeding. This may not therefore work in deeper or faster-running water, though some weight down the line (most of the weight in waggler fishing is typically bunched around the float) can result in the bait tripping the bed gently as the float processes downstream.

Pole fishing also offers a sensitive approach in static or gently flowing water. When using pole tactics, the float needs to be of a size appropriate to the current, with weights positioned to get the bait down to the feeding fish. If the fish are feeding closer to you, or can be drawn close in, a whip – a short pole – offers a method that not only is very sensitive but is also a means for swinging gudgeon to hand swiftly enough to enable match anglers to put together a substantial bag of fish.

Izaak Walton also notes:

> But many will fish for the Gudgeon by hand, with a running line upon the ground, without a cork, as a Trout is fished for: and it is an excellent way, if you have a gentle rod, and as gentle a hand.

Freeline fishing is indeed an efficient and rapid means to catch gudgeon when they are close by and feeding hard in shallow water. Sometimes, a shot placed close to the hook can give you a little additional casting weight and ensure that the bait reaches the bed quickly, with bites detected by watching for movement of the line, or sensing it with the finger tips.

When float or freeline fishing, groundbaiting can be important to both attract gudgeon and to keep them feeding. As noted earlier, some form of groundbait, proprietary or something as simple as liquidised or mashed bread, can be introduced to attract and hold the attentions of gudgeon. Whatever the type of groundbait, it should be mixed to a consistency appropriate to reach the bed. In still or very slow-flowing water, it is best to mix the groundbait loosely to create a rain of fine particles and a certain amount of cloudiness in the water. In heavier flows, a denser mix is necessary enabling the groundbait to sink faster and break up only at the river bed.

If the selected hook bait is different from the groundbait, for example, when you are presenting maggot or worm on the hook rather than breadflake when feeding liquidised bread, then mix a little maggot or chopped worm in with the groundbait.

Legering in all its forms is also a really effective means of catching gudgeon, putting the bait down on the river, canal or lake bed where the gudgeon live. This may be bomb fishing – presenting a light running leger with some groundbait in the swim – or better still using a swimfeeder to introduce feed directly into the feeding zone. I tend to use a cage feeder to introduce bread or other similar feed into the swim, or some other form of closed swimfeeder to introduce maggots, casters, chopped worm or other bait if fishing with these as hook baits.

I have not thus far mentioned line strength and hook size. Line strength is not critical for gudgeon fishing. However, if you are setting out your stall for gudgeon, solely or among other small fishes, light line will offer you optimum sensitivity and presentation. Hook size is generally small, size 22 to 18 ideal for baits such as maggots or larger hooks up to size 16 or 14 for bigger baits such as small worms and breadflake or punched bread.

In truth, gudgeon are a relatively easy fish to catch using a range of tactics, whether for pleasure or as a target for match anglers. As Izaak Walton noted, earlier, of the gudgeon:

He is an excellent fish to enter a young angler.

Mr and Mrs S.C. Hall describe in their 1859 book *The Book of the Thames From Its Rise to Its Fall*, albeit in rather patronising terms:

Gudgeons swim in shoals, are always greedy biters, and a very small degree of skill is therefore requisite to catch them; it is the amusement of ladies and boys more frequently than of men . . .

Gudgeon are, in fact, wonderful targets for anglers of all abilities and ages. Even most specialist and specimen anglers are happy to see a gudgeon, as evidenced by the many words written about them.

LURE FISHING FOR GUDGEON

For most of us, lure fishing is something we associate with predatory fish such as pike, perch, zander and trout, including opportunistic predators such as chub.

However, most, if not all, freshwater fishes, other than dedicated predators, tend to be opportunistic rather than entirely specialised in diet. Dace, chub and roach, for example, can be targeted effectively with artificial flies imitating fry or large invertebrates. Other fish species too, particularly as juveniles, prey in smaller animals such as water fleas, small insects and worms. Common and silver bream and many other similar species feed extensively on bloodworms, as well as a range of other similar worms and grubs.

My angling friend Richard Widdowson is an expert on the use of tiny lures, an inch or so and sometimes less in length with equally small hooks. He targets and catches gudgeon as well as a huge diversity of other freshwater and sea fishes not normally associated with lure fishing. From tench, roach and carp to eels and ruffe, Richard has caught freshwater fish of many species using these tiny lures, either on small jig heads or by dropshotting.

For jig fishing, small hooks (sizes 12 or 10) with weighted heads and an offset eye have a rubber worm or similar lure threaded on to them. These are then jigged along the bed of the waterbody, or against any walls, posts or other structures.

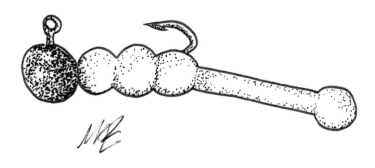

Dropshotting entails the lure being threaded onto a hook tied up the line using a palomar knot, such that the hook stands out at 90 degrees. The line is kept taut, and the lure is maintained at the right depth by a weight connected to the end of the line. Dropshotting offers very delicate presentation with a largely still or slowly moving lure. It also enables the use of much smaller hooks, or larger hooks if desired. Special weights for dropshotting with sprung clips are available, readily moved up or down the line to vary the depth at which the lure is presented.

Taking off the blinkers of entrenched angling habits and tactics, it should be no surprise that fishes of all species will take small lures imitating elements of their natural diets. Richard has had great success with gudgeon and other species, recommending red micro-lures in particular. I hope to emulate at least some of his success when I try this technique in warmer weather!

GUDGEON RECORDS

The British Record at the time of writing – a hostage to fortune as the record may be broken any day! – is held by a magnificent fish weighing 5 ounces, caught by D.H. Hull in 1990 from the River Nadder (a tributary of the Hampshire Avon) at Sutton Mandeville, Wiltshire.

D.H. Hull's momentous fish displaced the former record, a gudgeon of 4 ounces and 4 drams taken by M. Bowen in 1977 from Ebbw Vale Pond, Ebbw Vale, Blaenau Gwent, Wales. M. Bowen's fish in turn had displaced the previous record of 4 ounces taken by M. Morris in 1971 from Susworth Roach Ponds, Susworth, Lincolnshire.

Many older anglers will recall the purge in 1968 of many pre-existing British Record fish by the British Record (Rod Caught) Fish Committee (BRFC) due to lack of creditable evidence. Despite how we may feel about rejecting yesterday's records on the basis of their failure to meet modern standards, three gudgeon of 4 ounces and 4 drams were swept from the record books. The record holder to that date was George Cedrick for a gudgeon caught in August 1933 from the River Thames at Datchet in Berkshire. This fish was equalled by W.R. Bostock in October 1935 from Hoggs Pond in Shipley, Derbyshire, and again by J.D. Lewtin in 1950 from the River Soar in Leicestershire.

Delving back even further into history, Eric Marshall-Hardy notes in his 1943 book *Coarse Fish* four other notable gudgeon each weighing 4 ounces caught, respectively, by Mr R.G. Streeter from the Kentish River Stour in March 1932; Mr L. Reed from the River Tees in February 1935; Mr W.O. Limm at Tamworth in July 1935; and Mr G. Morritt from the River Medway in February 1936.

At the time of writing – another hostage to fortune! – the Scottish gudgeon record is unfilled but open to claim at 3 ounces and 8 drams.

However, leviathans of the gudgeon world though these fish may be, larger fish way beyond these sizes have been written about.

'Omar', in the 12 July 1958 edition of The Fishing Gazette, cited 12 gudgeon caught between 1951 and 1956 that exceeded the then-existing record.

Eric Marshall-Hardy, then-editor of Angling magazine, wrote in his delightful little 1943 book Coarse Fish about four gudgeon each weighing half-a-pound found when a reservoir in Bungay, Suffolk, was drained down. Marshall-Hardy records an extensive string of correspondence corroborating that these fish were gudgeon rather than small barbel, though there is no record of how accurate weighing was undertaken.

Another truly gigantic gudgeon was reportedly observed by Richard Walker in the infamous Redmire Pool in Herefordshire. In his classic 1986 book Casting at the Sun, Chris Yates writes:

> Richard Walker recalls seeing a gudgeon that looked all of a pound, and that's four times heavier than the gudgeon record. I never saw anything like that even though I used to see a lot of gudgeon.

Scientific records (see the 1992 book Freshwater Fishes of the British Isles by Maitland and Campbell) also suggest that gudgeon can grow up to 220 grams (nearly 8 ounces) across their wider geographical range, if not in Britain.

A concluding thought about monster gudgeon is that this is a fish that looks very big for its weight. Like grayling, which look enormous at a given weight relative to many other species of fish, gudgeon too seem to have a low density! Many are the 'huge' gudgeon that, on being put on the scales, surprise their captors by weighing in at only a couple of ounces!

GUDGEON MATCH CATCHES

When gudgeon are feeding in a match angler's swim, or have been induced to feed, they can be a prime target for putting together a bag that may also include a few 'bonus fish', such as interloping roach, bream or other larger species.

The expert match angler perfects a rapid approach. This generally entails drawing the gudgeon closer through targeted feeding, presenting the bait with a short line, be that on a rod, pole or whip, enabling hooked fish to be swung rapidly to hand. A large bag of gudgeon into double figures (over 10 lb) is possible if these fish can be induced to feed hard.

Gudgeon can also be the match angler's friend and favoured target species in more challenging conditions, such as after a hard frost or a long cold spell that puts most other fishes off the feed. Matches can be won by anglers adept at scrabbling around for a few gudgeon, perhaps with the occasional 'bonus fish', as these are fish that can generally be relied upon to feed when other species are harder to induce.

GUDGEON AS BAIT

Gudgeon are frequently recommended as a perfect live bait for perch. Whatever your attitude to using live fish as bait for predatory species, there is no doubt that the hardy and active nature and the pearlescent coloration of gudgeon is highly attractive to perch. Pike too, as well as zander, are also tempted by a gudgeon live bait. Eric Marshall-Hardy also notes that small gudgeon are useful as a live bait for chub.

Gudgeon live baits are said to work equally well in still and running waters where they are present. Gudgeon live baits may be presented by freeline tactics, under free-roving or static floats, or running leger. The unfortunate gudgeon are lip-hooked either with a large single hook or with a light snap tackle of paired treble hooks.

Perch and Gudgeon.

Perch are notoriously sensitive to resistance, so many perch anglers do not use a wire trace. However, a trace should always be used if there is a risk of leaving hooks in a pike, which can readily bite through monofilament lines. A critical element of good perch fishing though is to minimise resistance, whatever the method and presentation.

Of the use of gudgeon as live bait, Izaak Walton wrote in *The Compleat Angler*:

> *Take a small Bleak, or Roach, or Gudgeon, and bait it; and set it, alive, among your rods, two feet deep from the cork, with a little red worm on the point of the hook: then take a few crumbs of white bread, or some of the ground-bait, and sprinkle it gently amongst your rods. If Mr. Pike be there, then the little fish will skip out of the water at his appearance, but the live-set bait is sure to be taken.*

The use of gudgeon as live bait appeared to be popular in Victorian England, as evidenced by this extract from the chapter Gudgeon Fishing in the 1875 book Life on the Upper Thames by H.R. Robertson:

> Gudgeon are much used as bait when trolling for jack, and as a live bait for various large fish.

In relation to the use of gudgeon as dead bait, Walton wrote:

> And for your DEAD-BAIT for a Pike: for that you may be taught by one day's going a-fishing with me, or any other body that fishes for him; for the baiting your hook with a dead gudgeon or a roach, and moving it up and down the water, is too easy a thing to take up any time to direct you to do it.

What Walton describes about moving a dead bait 'up and down' is what we now refer to as 'sink-and-draw', animating the dead fish to mimic an injured prey fish to attract the attention of a predator. It is an effective method for covering a wide area of water.

Gudgeon may also be used effectively as a dead bait for various predatory fishes such as pike, zander and eels, as well as chub, and such opportunists are barbel.

Dead gudgeon have also been used for the once-popular practice of spinning in mounts, particularly when targeting game fishes. As Mr and Mrs S.C. Hall describe of trout fishing on the Thames in their 1859 book The Book of the Thames From Its Rise to Its Fall,

> the 'fish of size' are usually caught by 'spinning', the bait being a bleak, a small dace, a gudgeon, or a minnow.

Although gudgeon can be caught as bait fish by rod-and-line, another effective way to obtain numbers of gudgeon is by use of a bait trap. A standard minnow trap can be used for this purpose. Gudgeon can also be caught by cast nets, as described in the chapter *Gudgeon Fishing* in the 1875 book *Life on the Upper Thames* by H.R. Robertson:

> *When the fisherman requires them for this purpose, he seldom has recourse to the rod and line, but employs the casting-net, which soon supplies him with as many as he wants.*

However, a licence may be required to undertake this activity.

GUDGEON AS PROBLEMS!

For fans of gudgeon, it may sound like heresy to hear that there are anglers out there that regard gudgeon as a problem!

These are generally anglers seeking bigger species – carp, tench, bream and others – expressing their frustration that prolific gudgeon stocks get to the bait first.

Three principal methods may be deployed to avoid being swamped by gudgeon. Firstly, fish the bait off the bottom, be that under a float or 'popped up' as a buoyant offering wafting in the water column above the leger weight. Secondly, use a bigger bait than a gudgeon can fit in its mouth such as larger grains of sweetcorn, boilies (boiled baits) or lumps of paste baits. Thirdly, allied to the second method, use a harder bait that smaller fishes, including gudgeon, cannot readily nibble away. This third consideration is one of the reasons for innovation of hard-skinned boilie baits, but also bear in mind that there are many plastic imitations of 'real' baits that are now readily available that serve this purpose as well or better.

However, I assume that no-one reading these pages will regard catching gudgeon as a problem!

GUDGEON: THE PEOPLE'S FISH!

Many or most anglers, from novice to expert, simply love the humble gudgeon!

Izaak Walton wrote in *The Compleat Angler*:

He is an excellent fish to enter a young angler, being easy to be taken with a small red worm, on or very near to the ground. He is one of those leather-mouthed fish that has his teeth in his throat, and will hardly be lost off from the hook if he be once stricken.

The obliging gudgeon is, indeed, the first, or among the first, fish caught by many a novice angler. As Chris Yates related of his very earliest angling experiences in the 1993 book *A Passion for Angling*:

Then, acting on a bit of invaluable advice, I eventually caught my very first fish – a glorious gudgeon at least four inches long. It was a wonderful moment and also a fateful one: suddenly, I was a real angler.

This is an experience that many of us share!

Furthermore, in his classic 1986 book *Casting at the Sun*, Chris Yates writes of fishing Redmire Pool in Herefordshire:

> There were times when, frustrated by the carp, I would sit on the dam and console myself with a bit of tiddler snatching. Gudgeon fishing is the perfect antidote to the emotional pressures of stalking big pernickety carp. At Redmire especially, it seemed a particularly silly thing to do, which added to the enjoyment. There I was, a confirmed carp-addict, blessed with the honour of being able to fish the greatest carp water in the land. And what do I do? I fish for gudgeon. How irreverent!

Speaking with Chris Yates when I was writing this book, I found that his passion for the gudgeon is undiminished by the passing of years. In fact, he was really pleased to hear that, at long last, this wonderful little fish was about to have a book dedicated entirely to it!

I, of course, love gudgeon too. Mainly, they come along as a by-catch when I am fishing for roach or dace, though I do from time-to-time fish for them by design.

When fishing for them deliberately, the biggest problem I usually encounter is keeping the bait in the water long enough without it attracting the attention of minnows!

As we have observed, many or most anglers from novice to expert have a surprisingly deep affection for this little fish: the angler's favourite tiddler!

Four

Gudgeon are fascinating little fishes from biological and piscatorial points of view. However, they have far wider meanings
and cultural associations. In this chapter, we explore some
of these broader associations between gudgeon and people.

THE CHARM OF THE GUDGEON

I don't know what it is about gudgeon that always makes me
smile. In this regard, I am far from alone among the many
anglers – 'grown-up' and youngster alike – who profess an
irrational love of these cheery little fish.

It is understood that gudgeon can swell the nets of
match anglers. However, a more profound yet simplistic
pleasure is shared by many of us for this cheery little
fish, regardless of its size. As Chris Yates narrated, he even
enjoyed intervals catching them in legendary big carp
mecca Redmire Pool.

Chris is very far from alone among specimen anglers
who enjoy the odd foray after the humble gudgeon in other

DOI: 10.1201/9781003311010-4

esteemed 'big fish' waters, as well as for a welcome break from concentrating on larger quarry.

Like many river anglers, I have often halted my pursuit of heftier specimens when distracted by the glorious sight of a shoal of gudgeon, heads down and grazing together like sheep, on a shingle run beneath my rod tip. I need not even fish for them, though I may often enjoy a brief interlude doing so; simply watching the antics of these fascinating little fishes can be pleasure enough.

Furthermore, even on the coldest of days, we can rely on a gonk to sample our baits, brightening up chilly and slow sport, or even saving a blank, when trotting or legering for roach, dace or chub.

For me, as for many other people, gudgeon are also part of the background of a childhood spent by running water. Initially with cotton or a strand of monofilament tied to a whip cut from a hedge, and later with 'grown up' fishing tackle, gudgeon were a welcome catch throughout my younger years. This fascination, and also their demise in some of the places I knew, also pushed me towards my environmental work throughout later life. I still recall the sadness and anger welling up in my young heart as I saw a beloved stretch of the River Medway in Kent, a paradise where I cut my angling teeth, polluted and literally foaming in the early 1960s and then further neutered by canalisation for insensitive and myopic flood management and land drainage in the late 1960s.

I have caught many gudgeon since, and from many places, but perhaps that childhood link is part of the magic of these little fish. I suspect that this is exacerbated for many or most of us by the psychological impact of their almost cartoon-like bulbous heads and big eyes.

But I know that I am far from alone in loving the humble gudgeon, those abundant little river jewels eager to take crudely presented baits in any season!

GUDGEON ETYMOLOGY

The modern English word 'gudgeon' derives from Middle English 'gojoun', which originated from the Middle French word *goujon*. The first recorded use of the term 'gudgeon' was in the fifteenth century. The Latin word *Gobio* itself means 'gudgeon'.

As gudgeon are widespread and popular, it is perhaps unsurprising that they also go by a range of nicknames. They are widely and affectionately known as 'gonks'. 'Gobies' is another common name based on their Latin name *Gobio gobio*. At least a few people, with tongue firmly in cheek, refer to Welsh gudgeon as 'goboyos'. In the Midlands, the name 'pongo' is not uncommon.

Gudgeon are often also known as 'Trent barbel', or at least they were until the Trent become a premier barbel river after its recovery from industrial pollution! Elsewhere in the country, gudgeon are often given related local names, such as 'Suffolk barbel'. The subtitle of the chapter on gudgeon in my 2008 book *The Little Book of Little Fishes* is 'The poor man's barbel'.

Another nice, localised name for gudgeon is 'plummets', as you can keep adjusting your float until you catch one. This confirms that you have reached the bed of the river!

The Latin name *Gobio gobio* has long been accepted in science. However, over past centuries, various other Latin names have been applied to this fish. These include (among many others): *Cyprinus gobio* (the original name applied by the taxonomist Linnaeus in 1758); *Leuciscus gobio*; *Gobio fluviatilis*; and *Gobio vulgaris*. All now are superseded by the modern name *G. gobio*.

Gudgeon also go by a range of common names in the languages used across their wide geographical range. These include the following:

French	Goujon
Dutch	Grondel
German	Grundling
Danish	Grundling
Swedish	Sandkrypare
Italian	Ghiozzo
Spanish	Albur
Polish	Kiełb, Kiełb pospolity
Russian	Пескарь (Peskar)

Izaak Walton wrote in *The Compleat Angler*:

> *The Germans call him Groundling, by reason of his feeding on the ground; and he there feasts himself, in sharp streams and on the gravel.*

Whether this is true or not, I cannot say definitively, though I have my doubts. After all, the German word for 'ground' is actually *Boden*!

The word *goujon* – the French word for 'gudgeon' – is commonly used to refer to strips of fish, chicken breast or other meat, deep-fried coated in breadcrumbs. It is generally believed that the shape of the fish inspired the culinary *goujon*, also known by a diversity of local names such as chicken fingers, chicken tenders, chicken strips and chicken fillets. However, *gudgeon goujons* do not appear on any menu, as far as I know, perhaps as they are already *goujon*-shaped!

'Goujon' is also the name of a French automobile, manufactured between 1896 and 1901. This automotive Goujon was a four-seater voiturette (a miniature automobile) featuring an air-cooled, 3½ horse-power engine.

Two American submarines have also gone by the name 'Gudgeon'. USS Gudgeon SS-211, a long-range Tambor-class vessel commissioned in 1941, was the first American submarine to sink an enemy warship in the Second World War notching up 14 confirmed kills before being presumed lost in 1944. The second ship of the United States Navy to bear that name was a Tang-class submarine, commissioned in 1952 and eventually sold to Turkey in 1983.

The word 'gudgeon' also relates to socket-like, cylindrical fittings forming the female element of a hinge into which the male element (the pintle) locates to provide a pivot. This kind of hinge is commonly used for shutters and boat rudders.

Perhaps related to the fact that the capture of gudgeon has generally been regarded as simple, Eric Marshall-Hardy writes in his 1943 book *Coarse Fish*:

> the verb 'to gudgeon' crept into the English language in the 18th century, meaning roughly, 'to make a fool of by deception'.

However, as we will see later, when considering the contribution of the gudgeon to literature, the use of the term to denote foolishness predates this by centuries.

GUDGEON AND THE CREATIVE ARTS

Beyond the many biological and piscatorial references cited throughout this book, the humble gudgeon has also featured in a range of other, more high-browed literature.

Although Eric Marshall-Hardy wrote that 'the verb "to gudgeon" crept into the English language in the 18th century' metaphorical of foolishness, the use of 'gudgeon' in this sense is considerably older.

No lesser a writer than William Shakespeare (he who fished with angling author John Dennys) uses the metaphor of the foolish gudgeon in his ᶜ1596 play *The Merchant of Venice*. Antonio, the merchant of the play's title, famously entered into a contract with moneylender Shylock, committing him to owe one pound of his flesh if the loan was not repaid within three months. The character Gratiano, Shylock's most vocal and insulting critic during the eventual trial, says to Antonio after stating 'Let me play the fool':

> I'll tell thee more of this another time.
> But fish not with this melancholy bait
> For this fool gudgeon, this opinion.

The poet and satirist Samuel Butler wrote in his mock-heroic narrative poem *Hudibras*, published in the seventeenth century in the aftermath of the English Civil War as a scathing satire of Puritanism and the Parliamentarian cause from a Royalist perspective:

> Make fools believe in their foreseeing
> Of things before they are in being
> To swallow gudgeon ere th'are catch'd;
> And count their chickens ere th'are hatch'd.

A more recent reference to our favourite tiddler is included in the philosophical novel *The Brothers Karamazov* by the Russian Fyodor Dostoyevsky, published in instalments from 1879 to

1880. In this, he wrote, more than a little cryptically (I have to admit baffling me entirely!):

> You save your souls here, eating cabbage, and think you are the righteous. You eat a gudgeon a day, and you think you bribe God with gudgeon.

Unlike trout, for which Franz Schubert wrote a lied (song) and quintet titled *Die Forelle*, or the Atlantic salmon for which Henry Williamson wrote the novel *Salar the Salmon*, I am unaware of gudgeon inspiring major works of musical creativity or entire books.

However, the subset of illustrations throughout this book reflects artistic inspiration provided throughout the centuries by this popular little fish. Many of the images used in this book are out-of-copyright, and some are by the author. There are many more gudgeon illustrations in publications on fish and fishing.

Among these artworks is the rather wonderful painting that appears on the book's cover by fish and wildlife artist David Miller. David's gudgeon painting was used on the 2018–19 Environment Agency two-rod coarse angling rod licence, reflecting the wide appeal of this small fish, and was enthusiastically received by the angling public.

GUDGEON CUISINE

And so, onto the topic of eating gudgeon!

Many are the writers of fish and fishing that have extolled the virtues of gudgeon as food. Izaak Walton wrote in *The Compleat Angler*:

> The GUDGEON is reputed a fish of excellent taste, and to be very wholesome.

Angling author 'BB' (Denys J. Watkins-Pitchford) penned the essay *Food for Men*, published in his classic 1945 *The Fisherman's Bedside Book*:

> The best eating of all coarse fish are gudgeon and perch.

The flesh of a gudgeon is said to be finely textured and, apparently, tasty too. As the fish is small, eight gudgeon per person is considered sufficient. To prepare the fish, the head is cut off and the guts cleaned out after which it is washed and dried. Thus prepared gudgeon can be fried in butter or oil after being rolled in flour, seasoned to taste with salt and pepper during cooking. They may also be deep fried.

Arthur P. Bell writes in his 1926 book *Fresh-Water Fishing for the Beginner*:

> A dish of gudgeon fried a la whitebait is well worth eating.

The Reverend W. Houghton too, in his 1879 book *British Fresh-Water Fishes*, commends the flavour of the gudgeon, stating:

> In point of flavour the Gudgeon approaches that of the Smelt or Sparling, and in my opinion is one of the best of fresh-water fish we possess.

To this, Mr and Mrs S.C. Hall describe in their 1859 book *The Book of the Thames From Its Rise to Its Fall*:

> *If people care to eat, as well as catch, fish, there is no fish of the Thames more 'palatable' than the gudgeon, fried with a plentiful supply of lard. It is 'of excellent taste, and very wholesome', and has been sometimes called 'the fresh-water smelt'.*

Richard Franck notes in his 1694 book *Northern Memoirs*:

> *As the gudgeon is a most delicious fish, so ought he to be most delicately drest.*

Henry Coxon, in his 1896 book *A Modern Treatise on Practical Coarse Fish Angling*, comments:

> *Gudgeon are not only prolific fish, but are excellent for the table.*

Indeed, the taste of gudgeon seems to have been appreciated by many, particularly from and preceding the Middle Ages through to post–Second World War Britain, when coarse fish were still often taken for the table.

A common Middle Age dish was gudgeon tansy, the cleaned fish cooked with the bitter waterside herb tansy (*Tanacetum vulgare*) in much the same way, though in all probability a simpler version than the minnow tansy described by Izaak Walton.

BB offers a more imaginative serving of gudgeon in a tale about an idyllic fisherman's inn, in his 1987 book *Fisherman's Folly*:

> *Soon came Mrs. Small bearing a tray loaded with plates, a crusty cottage loaf, fresh lettuce plucked a minute ago from the garden, a dish of freshly caught gudgeon fried in bread crumbs and laced with watercress, and a flap of small-cured ham on which was enthroned two 'half apricots' of fried eggs still sizzling! It was unbelievable!*

In *Small Fry and Bait Fishes: How to Catch Them*, Kenneth Mansfield is generous in his commendation of the gastronomic virtues of gudgeon, with a further interesting insight into social history:

> Gudgeon are fiddly little fish that take time to prepare so they have become unpopular, but they are well worth the trouble. Until the end of Edwardian times they were a much sought-after fish and were sold (usually alive) by most fishmongers.
>
> I strongly advise anyone who has not yet tasted gudgeon to give them a trial when next he catches a dozen or so. He will go back and try to catch a gross!

In his 1943 book *Coarse Fish*, Eric Marshall-Hardy writes in very complimentary ways about the gastronomic virtues of this little fish:

> There are, quite frankly, few coarse fishes which appeal to my palate, but Gudgeon are very definitely an exception. I would rather by far a fry of these delicious fish than a trout or grayling, and I have much experience of them all. Take your Gudgeon home alive in a bait-can. To prepare them for eating, first bring a pan of deep fat to the boil. With this handy, take a wooden board and a sharp knife and decapitate each fish, open the stomach, clean it with the thumb. Now roll the fish in seasoned flour and case each into the boiling fat for three minutes. Prepared in this way, eaten with a squeeze of lemon, a sprinkling of red pepper and brown bread and butter, Gudgeon will hold their own with the best of fish.

Gudgeon also often feature in 'la friture' (whitebait-style fried fish), virtually a national institution in France.

As a final recipe, I am indebted to Jonathan Parsons for directing me to a wonderfully entertaining literary descriptive recipe for them. This was written by the famous cookery author 'La Mazille' from his 1929 book *La Bonne Cuisine de Périgord*. Jonathan not only kindly supplied this to me in French but also with his translation as follows:

The good gudgeon of Périgord.

I will not tell you that they are better than anywhere else, but I will tell you only that they are of good size, fat and plump and quite abundant along certain pure streams that afford, from time to time, this delicate treat of beautiful fried gudgeon.

We fry them there as everywhere, but when they come out piping hot we sprinkle them lightly with vinegar and we cover them with a little parsley vinaigrette made with garlic or shallots.

This delicious dish is especially enhanced by the frank and fragrant hillside wines. The light white wine of the region, clear and sparkling, with a pleasant flinty flavour, washes down the gudgeon most appropriately.

When you are ready to fry them, you wash and drain the gudgeon on a towel, after gutting only the larger ones. You roll them alive on a floured plate[1], and you cruelly throw them into a pan full of boiling oil.

The poor gudgeon lead their supreme dance and they jump in the air in their last convulsions. Anyway, the heat kills them straight away.

You must be careful not to put too many at once in the pan, so that they become well golden and crispy.

As you go, you scoop them out with a skimmer and put them in a long dish that you keep warm. Season with fine salt, drizzle with vinegar or verjuice and sprinkle the fish with a little parsley vinaigrette, either made with shallots or garlic cut into thin strips'.
[1]Many cooks do not flour the gudgeon.

Poor old gudgeon! Not only do you 'cruelly throw them into a pan full of boiling oil', but, quite clearly, 'gutting only the larger ones' happens when they are still alive!

In this and other recipes, gudgeon appear to have suffered for their allegedly exquisite flavour down the years. Happily, we tend to largely leave these inquisitive and colourful fish to their own devices these days!

True life confession: I have never – at least at the time of writing nor to my knowledge – eaten gudgeon! I simply share these writings with you as a matter of interest and, perhaps, your own experimentation.

THE SOCIAL ROLE OF GUDGEON

In Victorian England, fishing for gudgeon was a popular social pursuit. Furthermore, it was one very much enjoyed by the upper echelons of society. People (usually lower-class ones!) would be employed to rake the bed of the river to attract these inquisitive fishes and to expose natural invertebrate food upon which they would freely graze. Parties of anglers would then fish for them, often from punts, using canes, fine lines and, usually, red worms for bait. Ghillies would position the boats on their ryepecks (stakes in the river bed to which boats were tied), bait the hooks and unhook the fishes for parties including both ladies and gentlemen. Large catches could be made, and these were then generally cooked up as a feast for the angling party.

Kenneth Mansfield writes of this social form of gudgeon fishing, in his 1958 book *Small Fry and Bait Fish: How to Catch Them*, noting:

> . . . in the 19th century it became a fashionable pastime on the Thames. Many people who never fished for anything else organised or took part in gudgeon-fishing expeditions on that river, hiring a punt and boatmen for the purpose. Amply supplied with food and drink, and with the professional Thames fisherman seeing to such practical matters as propelling the punt, adjusting the ryepecks, raking the bottom and baiting the hooks, the anglers, male and female, enjoyed in comfort the sport of gudgeon fishing.

Anglers of wider experience did not despise such expeditions, and many of the really famous fishermen of the day – Francis, Foster and Buckland among them – described the pleasure they gained from such convivial outings.

This picnic version of gudgeon fishing lost its popularity soon after the turn of the century and disappeared with many other pleasing idiocyncracies [sic] of the nineties. Real anglers pursued larger quarry and the little gudgeon was left to the young, the match fisherman, and the seeker after live bait.

The popularity of gudgeon angling parties and the quantity of gudgeon caught are celebrated in the chapter Gudgeon Fishing in the 1875 book Life on the Upper Thames. As its author, H.R. Robertson, writes:

Old anglers tell us that the gudgeon are on the decline in the Thames, both as to number and size. They 'remember the time' when eighty dozen were to be taken in the day by the party in one punt. Now, at the present time, in a take of fifteen or sixteen dozen, it is seldom a really sizable fish gets in the wells. If the extremity of the bye-laws of the fishery were carried out, every gudgeon fisher, as he carries away his fish, would be indictable for taking unsizable fish.

GUDGEON AND NATURE CONSERVATION

Many freshwater fish species are of direct nature conservation concern, key parts of the biodiversity of freshwater habitats that are regarded as among the most threatened ecosystems globally. Thirty-eight per cent of Europe's freshwater fish species are threatened with extinction, with a further 12 fish species already declared extinct. At a broader scale, approximately 20% of the world's 10,000 freshwater fish species are listed as threatened, endangered or extinct. Our concern for fish should be far more than altruistic: this decline indicates a commensurate reduction in the vitality of ecosystems, including their capacities to support our various needs into the future.

Unlike species such as arctic charr, Atlantic salmon and European eel with threatened and declining populations, or non-native species such as topmouth gudgeon and black bullhead that pose a conservation threat to native species and ecosystems, gudgeon are not a priority under various strands of nature conservation legislation. Under the 'Red List' (IUCN 'Red List of Threatened Species'), documenting extinction threat, gudgeon are listed as 'Least Concern' (LC). Neither the Bern Convention ('The Bern Convention on the Conservation of European Wildlife and Natural Habitats 1979'), the European Union (EU) Habitats Directive (Council Directive 92/43/EEC on the Conservation of Natural Habitats and of Wild Fauna and Flora) nor UK nature conservation legislation (for example, the Wildlife and Countryside Act 1981, as subsequently amended) lists gudgeon for any special protection. However, the Bern Convention does impose bans on a range of destructive fishing methods.

Gudgeon nevertheless remain important elements of complex food webs in fresh waters. Their vitality or demise should be a matter of concern, indicative of the wider health of ecosystems and the many benefits that our freshwater systems provide.

GUDGEON IN LEGEND

One of Aesop's famous fables from ancient Greece in the fifth century BCE is titled *The Dolphins, the Whales and the Gudgeon*. Translated into English, it reads:

> Some dolphins and some whales were engaged in battle. As the fight went on and became more desperate, a gudgeon poked his head above the surface of the water and tried to reconcile them. But one of the dolphins retorted: 'It is less humiliating for us to fight to the death between ourselves than to have you for a mediator'.

The meaning of the fable is that 'nobodies' (bystanders of no great importance) may end up thinking that they become a significant 'somebody' when they interfere in a public row. I can see strong parallels with contemporary media critics, and with those who do nothing creative yet define themselves by sniping at those who do!

There are, however, three problems with this tale. Firstly, Aesop was allegedly a slave, but there is considerable doubt that this was the case. Secondly, there is widespread speculation that Aesop may, in fact, be an entirely fictitious character. Thirdly, of course, our familiar gudgeon, G. *gobio*, is not naturally present in Greece. Regardless of the truth of these matters, the humble gudgeon is forever cemented in ancient legend!

PET GUDGEON

Gudgeon are one of the species of British freshwater fish that acclimatise quite readily to indoor aquaria, being hardy, not too large and unexacting in their water quality requirements. They are a fascinating fish to watch, and are happiest in small shoals. However, in my experience, you do have to keep the lighting subdued as they are easily spooked in bright or sparsely planted tanks.

Given their predilection to seeking out stray scraps of edible matter of all sorts from the bed of the river or pool, Kenneth Mansfield wrote:

> Some aquarium keepers like to have a gudgeon or two in their tanks to keep the bottom clean.

Some garden centres from time to time sell gudgeon for the home aquarist or pond keeper. Gudgeon can also be caught from the wild for this purpose, though only with the landowner's permission.

THE ECONOMICS OF GUDGEON

What is a fish worth? Well, you can buy a fish from a fishmonger, an aquarist supplier or from a purveyor of stock fish for the angling trade. But price alone is a poor surrogate for their many and differing values.

Economically, gudgeon may be important in the context of match fishing, sometimes constituting the staple catch in some match-fished waters. Also, as we have seen, they can be prized by pleasure anglers, youngsters for which they are often among the first fish caught, as well as attracting a new generation of lure anglers and others keen to explore formerly neglected small species. Gudgeon are also often used by anglers as baits for predatory fish species.

In France in particular, as in former times elsewhere, gudgeon are highly prized for their culinary worth with both commercial and cultural values, as explored with many examples earlier in this chapter.

Adding to their value to society are the various contributions of gudgeon to artistic creativity outlined previously in this chapter. This includes a wide range of drawings of the fish throughout the centuries, a small subset reproduced throughout these pages.

Gudgeon also constitute important links in food chains, feeding on a variety of small food items and in turn fed upon by many larger species of fish, water birds and some mammals too including otters. Take away the gudgeon, and these closely co-evolved ecological roles are lost at untold cost to the vitality of fresh waters and the many human uses and other values that they provide.

OTHER THINGS SAID ABOUT GUDGEON

Seemingly, very many people share a deep affection for the humble gudgeon!

In his marvellous little 1958 book *Small Fry and Bait Fishes: How to Catch Them*, Kenneth Mansfield states:

> *One of the many surprising things in the sport of angling is the seemingly disproportionate delight that anglers of all ages in all ages have taken in catching gudgeon; fish whose average weight is about 3 oz.*

Though their wonderful little 1925 book *Fishing: Its Cause, Treatment and Cure* was written with tongue planted firmly in cheek, H.T. Sheringham and G.E. Studdy betray their warm feelings for this delightful little fish:

> *This is the fish which 'Beauty draws with a single hair'. Persons who have the misfortune to be plain use a rake. The poet has beautifully said that the gudgeon rises 'to no fancy flies'. This is true as well as beautiful.*

The parallels with certain behaviours of the male of our species were not lost on John Gay (1685–1732), the English poet and dramatist best remembered for *The Beggar's Opera*. Of this kinship, and perhaps also reflecting some misfortune of his own, Gay penned the lines:

What gudgeons are we men,
Every woman's easy prey!
Though we've felt the hook, again
We bite and they betray.

Ausonius (Decimus Magnus Ausonius, c.310–c.395), the poet and man of letters from Gaul, wrote of the gudgeon too. Translated from Latin, he wrote:

Thou too, O Gudgeon, worthy of being mentioned among the shoals of the river, not greater than the two palms of the hands without the thumb; very fat, rounder and plumper still when they belly is full of eggs; Gudgeon imitating the hairs of the pendant-bearded Barbel.

GUDGEON SOCIETIES

People with interests in many larger fish species have formed societies dedicated to their study and pursuit as angling quarry. Pike, carp, barbel, tench, perch, chub, roach, dace and many more species besides have their societies, some with substantial membership and financial turnover. But the humble gudgeon too has its own bands of aficionados.

One such is The Gudgeon Society. Set up by Carl Smith, by his own admission as a kind of joke, the Facebook (social media) page for The Gudgeon Society at the time of writing (December 2020) has nearly 2,000 members on social media – including me! These are people sharing a passion for all things related to the gudgeon. The Gudgeon Society's logo,

appearing here, may look a little familiar from its appearance elsewhere in this book. After all, I designed it!

There is also the Grand Union Gobio Gobio Society (GUGGS), formed in 2009 with the wonderful motto 'Size doesn't matter . . .'. I fully concur! GUGGS members are dedicated to the pursuit and capture of canal gudgeon. The stated ultimate aim is for a GUGGS member to catch a specimen to break the UK gudgeon record, though generally they are happy to catch gudgeon of any size. Any gudgeon of 30 grams or more is determined to be a specimen fish that may be given a name. Rather wonderfully, GUGGS publishes on its website annual awards, GLAMs (GUGGS Ladies' Annual Medals), 'Gobio of the month', lists of specimen fish, and a guide to reaches of the Grand Union Canal with lists of the largest gudgeon caught there. Initially focused on the Grand Union Canal, GUGGS has since spread its wings to cover other canals where gudgeon can be caught.

Amazingly, these are not the only groups with specific interest in fishing for gudgeon.

Another like-minded gang of people with a passion for gudgeon is to be found among the Traditional Fisherman's Forum. The Forum comprises approximately 500 enthusiasts from across the UK, but also internationally with members across Europe as well as America, Canada and South Africa. All share an interest in traditional fishing tackle, and they meet periodically for good company, plenty of cakes and the odd bit of fishing.

The 'Jolly on the Wally' has been a regular event of the Traditional Angler's Forum since 2016, held on the bijou, delightful and mainly tree-covered River Wallington in south-east Hampshire. The fishing on the Wally is controlled by Portsmouth and District Angling Society, which fully supports these jollies. The river abounds with a range of fish species, even in reaches that can be practically stepped across in low flows. Gudgeon are the plentiful prime attraction of the 'Jolly on the Wally'.

Recognising my gudgeon-fancying tendencies, I was kindly invited to join the merrie gang on the Wally in late September 2021 for a 'cup match' (if such a laid-back event can fairly be called a match!). The loose assemblage of anglers scatters along the river to compete for the longest gudgeon, measured in inches against a ruler from snout to tail tip and backed up by photographic evidence.

The fates were with me on the day: I managed to catch a champion gudgeon measuring 6¾ inches from snout to tail (I had three that long!) taken on red maggots under a stick

float, using a 1950s' Rapidex centrepin and a glass fibre rod much refurbished since I bought it second hand in 1971.

For this, I was awarded The Gudgeon Jim Trophy, handed to me by the 2020 winner.

The Gudgeon Jim Trophy has been contested since 2017 in memory of the passing of Forum member 'Gudgeon Jim', so named imaginatively as he was called Jim and used to fish for gudgeon. I look forward to joining this merry gang again on the Wally to defend the gudgeon crown and to eat more cakes!

It is amazing how such a little fish can cause such a large amount of passion among ostensibly 'grown-up' people!

A GUDGEON STORY

In collating interesting snippets for this book, a real-life vicar – Jon Barrett – kindly shared with me the tale of *The Bishop, the Vicar, the Bishop's daughter and the gudgeon.*

This real-life tale relates to an eighteenth-century vicar. George Harvest (1715–1780) was Reverend of Thames Ditton, a preacher, eccentric and author. Here is the tale, as told to me by Jon:

> *Reverend George Harvest was engaged to be married to the daughter of the Bishop of London.*
>
> *On the morning of the 'big day', the Vicar decided to while away the hours before the nuptial ceremony with a spot of fishing for gudgeon – his favourite quarry.*
>
> *However, so engrossed was he in his angling activity that he missed the wedding.*
>
> *His intended, clearly not herself an angler, took umbrage and broke off the union before it had another chance to be formalised or blessed.*

Jon Barrett told me, speaking as an angler rather than as a representative of the established church, that he thought that The Reverend Harvest had a lucky escape. After all, the Bishop's daughter would surely never have understood the 'one last cast' mentality that permanently leads virtually all of us anglers to arrive home later than promised!

We all have our gudgeon stories. This book is dedicated to all those who love this cheery little fish and the stories we can tell about it!

Gudgeon Bibliography

The following works are referenced in this book, with my thanks to the authors concerned where quoted.

'BB' (Watkins-Pitchford, Denys J.). (1987). *Fisherman's Folly* (Revised edition.). Boydell Press, Woodbridge (Suffolk).

'BB' (Watkins-Pitchford, Denys J.). (1993). *The Fisherman's Bedside Book* (Revised edition.). White Lion Books, Cambridge (Reprinted by Merlin Unwin Books, Ludlow, 2004).

Bell, Arthur P. (1926). *Fresh-Water Fishing for the Beginner.* Warne's Recreation Books, Frederick Warne & Co. Ltd., London.

Berners, Dame Juliana. (1496). *Treatyse of Fysshynge with an Angle.* Wynkyn de Worde of Westminster, London.

Coxon, Henry. (1896). *A Modern Treatise on Practical Coarse Fish Angling: How to Catch Fish.* Charles H. Richards, Nottingham (Republished in 2004 by The Medlar Press, Ellesmere).

Dennys, John. (1613). *Secrets of Angling.* Roger Jackson, London.

Everard, Mark. (2008). *The Little Book of Little Fishes.* The Medlar Press, Ellesmere.

Everard, Mark. (2020). *The Complex Lives of British Freshwater Fishes.* CRC/Taylor and Francis, London and Boca Raton.

Franck, Richard. (1694). *Northern Memoirs.* Henry Mortclock, London.

Hall, Mr and Mrs S.C. (1859). *The Book of the Thames from Its Rise to Its Fall.* Alfred W. Bennett and Virtue & Co., London.

Houghton, MA, FLS, The Reverend W. (1879). *British Fresh-Water Fishes.* William Mackenzie, London (Note: This book has been reprinted over the decades by numerous publishers, for example by The Peerage Press, London, in 1981).

Maitland, P.S. and Campbell, R.N. (1992). *Freshwater Fishes of the British Isles.* HarperCollins Publishers, London.

Mansfield, Kenneth. (1958). *Small Fry and Bait Fishes: How to Catch Them.* Herbert Jenkins, London.

Marshall-Hardy, Eric. (1943). *Coarse Fish.* Herbert Jenkins, London.

'La Mazille' (Mallet-Maze, Andrée). (1929). *La Bonne Cuisine de Périgord.* Flammarion, Paris.

Pinder, A.C. (2001). *Keys to Larval and Juvenile Stages of Coarse Fishes from Fresh Waters in the British Isles.* Freshwater Biological Association Scientific Publications Volume 60. Freshwater Biological Association, Windermere.

Robertson, H.R. (1875). *Life on the Upper Thames.* Virtue, Spalding, & Co, London.

Sheringham, Hugh Tempest and Studdy, G.E. (1925). *Fishing: Its Cause, Treatment and Cure.* Philip Allan and Co., London.

Tate Regan, C. (1911). *The Freshwater Fishes of the British Isles.* Methuen and Co. Ltd., London.

Walton, Izaak and Cotton, Charles. (1653). *The Compleat Angler.* Maurice Clark, London (Available these days in many editions and from various publishers).

Wells, A. Lawrence. (1941). *The Observer's Book of Freshwater Fishes of the British Isles.* Frederick Warne and Co. Ltd, London.

Yates, Chris. (1986). *Casting at the Sun: Reflections of a Carp Fisher.* Pelham, Books, London (Republished in 1995 and 2005 by the Medlar Press, Ellesmere).

Yates, Chris, James, Bob and Miles, Hugh. (1993). *A Passion for Angling.* Merlin Unwin Books/BBC Books, Ludlow.

Professor Mark Everard is a scientist, author and broadcaster, working on water and ecosystems around the world. He also has an irrationally large interest in fish!

Mark is a passionate angler, getting out whenever he can after coarse, game and sea fish. He has an enviable track record of specimen fish with a particular passion for roach, dace and mahseer and, of course, gudgeon, and the whole river ecosystems that support them. He has long been a champion of the 'little fishes'.

Mark Everard is often referred to as 'Dr Redfin' in the angling press for his special passion for roach.

For more details about Mark and his work, see www.markeverard.co.uk.